YOUR KNOWLEDGE HAS VALUE

- We will publish your bachelor's and
  master's thesis, essays and papers

- Your own eBook and book -
  sold worldwide in all relevant shops

- Earn money with each sale

Upload your text at www.GRIN.com
and publish for free

GRIN

Bibliographic information published by the German National Library:

The German National Library lists this publication in the National Bibliography;
detailed bibliographic data are available on the Internet at http://dnb.dnb.de .

Imprint:

Copyright © 2018 GRIN Verlag
Print and binding: Books on Demand GmbH, Norderstedt Germany
ISBN: 9783668730519

This book at GRIN:

https://www.grin.com/document/428650

Leonard Kahungu

Factors Influencing Demand Side Management Strategies

GRIN Verlag

FACTORS INFLUENCING DEMAND SIDE MANAGEMENT STRATEGIES

BSc (Hons) Oil and Gas Management

MERR5030-Managing Energy Resources and Regulations

Contents

Executive Summary

The future of UK's energy industry is currently facing significant threats attributed to dwindling energy resources and escalating energy demand. Demand side management strategies can play a central role in facilitating the UK to smoothly transition from relying on conventional energy production systems to modernised systems that are more reliable due to reduced production costs and enhanced yields (Torstensson & Wallin 2015). The DSM strategies are predominantly divided into three types namely demand side reaction, distributed generation, and demand reduction.

Demand-side responses are concerned mainly with short-term actions characterising the consumers in attempts to modify their power consumption patterns. Distributed generation seeks to establish localized energy generation and distribution to offset the demand load in macro grid power systems. Demand reduction is caused by escalating population growth and the increase of technological appliances and seeks to reduce the energy stress in macro grid power networks. Reducing energy demand explores significant ways to influence consumer behaviours and to enhance power consumption efficiency (Owens & Driffill 2010).

Each of DSM groups is affected by technological, financial, and institutional factors. Financial aspects explore monetary benefits and barriers while technological factors deal with technical potential and challenges inhibiting successful implementation of DSM strategies. The institutional elements mostly focus on regulatory issues influencing demand-side management strategies. Thus, the primary recommendations in the UK's context are concerned with technical developments and policy reconstitution to facilitate secure and sustainable energy management.

Factors Influencing Demand Side Management Strategies

The United Kingdom is facing the currently undergoing a trial moment due to growing concerns regarding the future of its energy system. More than two-thirds of the UK's power stations depend on old coal, gas, or nuclear-powered stations, which are expected to close by the year 2030. The closure of these stations is attributed to regional treaties such as the European Large Combustion Plant Directive, which compares the European Union signatories to honour clean air obligations (Torriti, Hassan, & Leach 2010). Besides, the existing energy system in the UK is incapable of achieving growing power demand and sustaining marketing requirements. This demonstrates that energy security, efficiency, and sustainability are some of the critical priorities in the UK's energy system.

Demand Side Management

Demand-side management (DSM) entails a series of strategic plans designed to reduce electricity or energy use through innovative programs and activities in efforts to promote energy efficiency and conservation, mainly through efficient energy management. It involves the planning, execution, and assessment of energy utilities targeting the consumers in attempts to promote efficient energy consumption and demand patterns (Torstensson & Wallin 2015). Therefore, effective DSM strategies are paramount in facilitating the accomplishment of secure, sustainable energy system in the UK. These strategies are particularly relevant to the future UK energy system as they are likely to facilitate a smooth transition from fossil fuel-dependent power stations to modernised energy generation, conservation, consumption, and distribution techniques. The UK's future energy system is more likely to benefit from DSM strategies in the midst of growing energy demands and declining fossil fuel resources in the energy system (Domínguez et al. 2012). Consequently, three major types of DMS strategies include demand-side response, distributed generation, and demand reduction, in which each of these categories has financial, technological, and institutional aspects.

Types of Demand Side Management Strategies

1. Demand Side Response

Demand side response also referred to as demand response (DR) are programs designed to encourage end-user consumers to enact short-lived energy demand in reaction to certain variables regulated by power grid operator or price signals from usual market ratings (Leeflang & Wittink 2013). The action is deemed to last from one to four hours and includes reduction of power consumption such as through turning off unnecessary lighting, dimming unnecessary lighting, shutting down complementary lighting or manufacturing processes and turning off superfluous residential, commercial or industrial components in attempts to off-load power demand from the power grid operator. These activities may also include using the onsite generator to displace the demand burden placed on the power grid (Segu 2012). DR encourages energy consumers to reduce the amount of energy consumed from their power grid at a specific time in reaction to predetermined indicators.

1.1. Financial Aspects

1.1.1. Drivers

Demand response is mostly concerned with monetary savings for both utilities and consumers. In other words, these strategies provide the energy market with monetary incentives in efforts to promote reduced or regulated energy consumption. Thus, the return on investment for both the consumers and DR utilities is one of the most fundamental aspects of this scenario. Policies and regulatory frameworks influencing the market charges and return on investment in DR utilities is likely to impact the reliability of managing demand side patterns (Kemp, 2011).

1.1.2. Barriers

Residential markets largely remain untapped which inhibits the acceleration of DSM strategies focusing on DR. Price incentives aiming at demand side response may be unsustainable due to market forces. This obscures the development of a structured market aiming to the demand-side reaction. Apparently, increasing financial incentives for the consumers do not always translate to reduced energy demand, restricting potential DR benefits in DSM.

1.2. Technological Aspects

1.2.1. Drivers

Numerous DR technical applications are designed to offset the energy demand from conventional power stations. On the other hand, solar systems and other renewable energies aiming to reduce the demand load are characterised by unreliable energy supply. Therefore, a breakthrough in efficient, reliable, and efficient storage technology is one of the critical drivers in DR approaches (Arteconi, Hewitt, & Polonara 2012). Cost-effective storage technologies are some of the major factors facilitating a faster integration of DR into the contemporary micro and macro power grids.

1.2.2. Barriers

Most DR utilities are essential in mitigating power uncertainties in macro-grids but cannot provide long-term technical solutions. Therefore, system reliability is a significant concern in the management of demand-side response and may affect consumer behaviours in adopting conservative energy behaviours (Pholboon, Sumner, & Kounnos 2016). In other words, DR utilities are designed to offer short-term demand solutions, and hence incapable of facilitating the achievement of secure, sustainable energy demand and supply.

1.3.Institutional Aspect

1.3.1. Drivers

The availability of legislative policies and regulatory frameworks that directly facilitate the implementation of DR utilities is fundamental in DSM. Such legislation and regulations create a favourable environment for the execution of DR plans, which is fundamental in the acceleration of demand-side reaction.

1.3.2. Barriers

Regulatory uncertainties are fatal to the full realisation of DR approaches. This is attributed to the confusion caused by political interferences, hindering the adoption of healthy conservative behaviours in the market. Moreover, the presence of ambiguous policy clauses is likely to discourage the exploitation of demand response utilities.

2. Distributed Generation

Distributed generation (DG) is used to describe decentralized or localized energy generation and storage through small-scale combined heat powering systems that include renewable sources of energy. DG connections are different from traditional power stations, including hydroelectric resources and large solar power stations. Most DG systems are based on small grid connections and often produce up to 50 megawatts (Drude Junior, & Rüther 2014). The small grid can easily be disconnected from macro grid systems to operate independently. DG systems are expected to have a central role in the UK's future energy security and management. These facilities will facilitate the integration of renewable resources into macro-grid methods to enhance the effectiveness of demand-side management (Vardakas, Zorba & Verikoukis 2015). Nevertheless, UK's ability to exploit distributed energy generation resources may be enhanced or hindered by economic, technical, and institutional factors.

2.1. Financial Aspects

2.1.1. Drivers

The installation, maintenance, and production costs of simplified solar power panels has declined in the last few years with more than 50% due to increased awareness and intensified conceptualisation of micro-grid power stations. Moreover, the intensification of micro-turbines and biomass power generation is increasingly becoming cheaper, promoting the integration of distributed energy resources to derive maximum benefits are relatively lower prices.

2.1.2. Barriers

Reliable power supply obtained from DG utilities interprets to an uninterrupted power supply to the consumers. Nonetheless, this necessitates regular maintenance, recalibration, and installation of advanced DG systems, which are relatively expensive to common consumers. Additional charges are resulting from contractual issues, legal and compliance requirements also hinder efficient installation of DG utilities, especially in small-scale consumers. The initial cost of installation is also relatively prohibitive, which hinders the acceleration of DG and microgrid power station development (Haney et al. 2010).

2.2. Technological Aspects

2.2.1. Drivers

The adoption of homogenous technical requirements in the UK is likely to enhance the interconnectedness of the DG power grid Universal DG technical tools encourages efficient integration and networking of micro-power grid systems into the macro-power grid infrastructures (Roldán-Blay et al. 2017). These measures are paramount in accelerating the development of DG planning, implementation, control, and monitoring expertise and systems, increasing the efficiency of DG utilities in addressing the existing energy demand and supply issues facing the UK.

2.2.2. Barriers

Different utilities play a fundamental role in the maintenance of safe and reliable DG grid systems. However, the resulting integration of small grid systems into the existing infrastructures is prone to conflicts of interest due to legitimate incentives to discourage self-power generation by the consumers. The existing distribution configuration systems ate less compatible with energy consumption utilities, which necessitates the utilization of numerous technical tools to conform to technical aspects of DG.

2.3. Institutional Aspect

2.3.1. Drivers

The contemporary regulations, policies, and legislation have started appreciating the significance of simplified and homogeneous contracts for small organizations promoting or striving to capitalise on DG utilities in the European Union region. The existence of these regulatory frameworks is likely to address prohibitive fee and compliance issues such as additional insurance liability, unilateral indemnification along with limitations on transfer of sales. The development of friendly DG regulations is likely to ease the installation of small-grid power generators, which will offset substantial demand load from the conventional power stations as the reliance on renewable sources continue to flourish (Koliou et al. 2015). This is likely to enhance the UK's ability to maximise the benefits of distributed generation in efforts to achieve sustainable and secure energy demand and supply.

2.3.2. Barriers

On the other hand, the present regulatory and utility guidelines comprise the major barriers inhibiting the implementation of efficient distributed generation in the UK. The access to wholesale markets, unbundled distribution, and an exclusive franchise is limited, which inhibits the ability to use and install DG sources among the consumers (Corbett 2013). Policies influencing tariffs aspects are often restrictive since the consumers are expected to comply with numerous charges and legal requirements, making DG acquisition for small-scale consumers an exorbitant affair.

3. Demand Reduction

There has been a significant increase in demand for energy, which is attributed to the intensification of population growth and major demographic changes in the UK and other parts of the globe. The escalation of electronic devices, technological development, and emerging economies are some of the factors influencing energy demand (Owens & Driffill 2010). This necessitates the implementation of strategic energy management approaches targeting demand reduction through various approaches such as smart constructions and upgrades, HVAC enhancement, re-commissioning and variable frequency drives. Studies indicate that the energy demand across the globe will grow with at least 2.2 annual percentage to 2020 with the existing environmental, socioeconomic and political conditions (Torstensson & Wallin 2015). Similarly, the UK is also expecting to record a considerable energy demand growth in the near future. These observations underscore the significance of energy demand reduction in the contemporary society.

3.1. Financial Aspects

3.1.1. Drivers

Price signals are some of the key determinants of energy consumption since they encourage the adoption of conservative power behaviours. Eliminating critical features that hide the actual cost of electricity from the consumers due to political interferences and interests is paramount in supporting a reduction of demand in residential and commercial environments. It is believed that consumers are likely to use conservative energy products and attitudes when they understand the real price of electricity (Domínguez et al. 2012). Thus, establishing robust metrics in carbon measurement and other resources facilitates the accurate billing in attempts to achieve demand reduction. Incentive aimed at reducing energy demand can also enhance the development of energy-efficiency appliances and buildings, together with the use of renewable energy sources.

3.1.2. Barriers

Economics in energy production and management comprise multifaceted variables that require critical and sensitive equilibrium. The failure to establish an equilibrium may trigger large-scale political and socioeconomic crisis and eventually damage the national, regional or international economy. Ideally, power is a universal requirement in virtually all economic activities (Leeflang & Wittink 2013). As a consequence, the inability to establish a balance between financial aspects and the need for demand reduction is likely to inhibit strategic measures put in place to lessen power demand load.

3.2.Technological Aspects

3.2.1. Drivers

Smart grid technology is an essential component in demand reduction as it allows consumers to appreciate and benefit from efficient appliances, particularly during peak periods. It allows residential and commercial environments to monitor energy consumption patterns and make reliable or informed decisions about consumption management.

3.2.2. Barriers

The installation of advanced technological appliances and innovative energy management techniques requires immense production capital, which may discourage tremendous investment in energy saving appliances and technical solutions (Owens & Driffill 2010). Maintaining the efficiency of energy-saving devices may also require an immense amount of resources, which may prohibit the use of these technical solutions.

3.3.Institutional Aspect

3.3.1. Drivers

Regulatory bodies evaluate the development of energy efficient electric appliances and buildings by prohibiting or penalising the production of inefficient appliances in power consumption. Significant institutions responsible for energy research and development is capable of influencing the development of highly efficient appliances, regulations, and policies favouring sustainable demand reduction strategies (Segu 2012). Thus, the government should encourage innovative methods to promote the culture of energy saving and establish tailored energy reduction tools aligning with geographical, commercial, social, and economic activities.

3.3.2. Barriers

Political interferences lead to the formulation of unreliable legislations that seek to protect consumer and market forces, by restricting the cost transferred to residential and commercial consumption billing systems. The development of policies that limit the transfer of some cost to end-users encourages power wastage behaviours since the society is protected from understanding the underlying implications of increased energy demand and consumption patterns (Arteconi, Hewitt, & Polonara 2012).

Conclusion

Apparently, the impending closure of coal and oil powered stations in combination with electricity demand pattern predictions are major concerns to the UK's future energy security and sustainability. The ultimate feasible solutions are to increase energy supply through the construction of efficient power infrastructures in attempts to increase electricity supply and managing the contemporary demand determinants, prior to the implementation of robust power generating infrastructures that comply with regional and international standards.

Particularly, DSM strategies are essential in ensuring that the UK's energy resources can meet the ever-growing energy demand by promoting consumption effectiveness. Three major types of DSM strategies include demand reduction, distributed generation, and lastly demand-side response. DSM categories are prone to financial, technological and institutional manipulation. Comprehending these factors is paramount in establishing strategic ways in which the United Kingdom can implement demand-side management in efforts to secure a sustainable and secure future for its energy resources.

Recommendations

A review of the contemporary factors influencing different types of DSM reveals major loopholes in the modern UK's energy resources warranting strategic recommendations. One of the major factors identified as the main barrier to DSM strategies entails institutional weaknesses. The UK is yet to put in place robust policy framework and policies that can enhance the efficiency of DSM utilities in efforts to lessen the existing energy load to traditional power stations relying on nuclear, oil, and gas resources. As a result, the UK ought to re-evaluate its energy regulations to enhance the distributed generation and demand reduction for power. These approaches are likely to cultivate an energy-saving culture among the consumers. Therefore, organisations dealing with the management of energy resources ought to collaborate in attempts to establish long-lasting solutions to the existing issues facing power resources, especially through the management of demand factor.

Additionally, there are significant gaps in technical development in DSM strategies. This is attributed to the inability to store energy, especially at the micro-grid level, which hinders the efficiency of distributed generation utilities. This calls for advanced and comprehensive research and development approaches aiming at increasing the efficiency of power storage. These activities should also focus on innovative techniques of promoting output efficiency with minimal costs and power consumption. This will facilitate the conservation of power and also enhance the effectiveness of exploiting the available energy resources in the UK. Besides, research and development technological activities will be beneficious to the UK's energy resources. This attributed to the assumptions that DSM strategies will facilitate a smooth transition from coal, nuclear, oil and gas powered-stations to modernised infrastructures in compliance with the global and regional requirements.

References

Arteconi, A., Hewitt, N.J. & Polonara, F., 2012. State of the art of thermal storage for demand-side management. *Applied Energy*, *93*, pp.371-389.

Corbett, J., 2013. Using information systems to improve energy efficiency: Do smart meters make a difference? *Information Systems Frontiers*, *15*(5), pp.747-760.

Domínguez, S., Sendra, J.J., León, A.L. and Esquivias, P.M., 2012. Towards energy demand reduction in social housing buildings: envelope system optimization strategies. *Energies*, *5*(7), pp.2263-2287.

Drude, L., Junior, L.C.P. & Rüther, R., 2014. Photovoltaics (PV) and electric vehicle-to-grid (V2G) strategies for peak demand reduction in urban regions in Brazil in a smart grid environment. *Renewable Energy*, *68*, pp.443-451.

Haney, A.B., Jamasb, T., Platchkov, L.M. & Pollitt, M.G., 2010. Demand-side management strategies and the residential sector: lessons from international experience. *The Future of Electricity Demand: Customers, Citizens and Loads.*

Kemp, I.C., 2011. *Pinch analysis and process integration: a user guide on process integration for the efficient use of energy.* Elsevier.

Koliou, E., Bartusch, C., Picciariello, A., Eklund, T., Söder, L. & Hakvoort, R.A., 2015. Quantifying distribution-system operators' economic incentives to promote residential demand response. *Utilities Policy*, *35*, pp.28-40.

Leeflang, P.S. & Wittink, D.R., 2013. Building models for marketing decisions: Past, present and future. *International journal of research in marketing*, *17*(2-3), pp.105-126.

Pholboon, S., Sumner, M. & Kounnos, P., 2016, October. Community power flow control for peak demand reduction and energy cost savings. In *PES Innovative Smart Grid Technologies Conference Europe (ISGT-Europe), 2016 IEEE* (pp. 1-5). IEEE.

Owens, S. & Driffill, L., 2010. How to change attitudes and behaviours in the context of energy. *Energy policy*, *36*(12), pp.4412-4418.

Roldán-Blay, C., Roldán-Porta, C., Peñalvo-López, E. & Escrivá-Escrivá, G., 2017, July. Optimal Energy Management of an Academic Building with Distributed Generation and Energy Storage Systems. In *IOP Conference Series: Earth and Environmental Science* (Vol. 78, No. 1, p. 012018). IOP Publishing.

Segu, R., 2012. Demand-Response Management of a District Cooling Plant of a Mixed Use City Development.

Torriti, J., Hassan, M.G. & Leach, M., 2010. Demand response experience in Europe: Policies, programmes and implementation. *Energy*, *35*(4), pp.1575-1583.

Torstensson, D. & Wallin, F., 2015. Potential and barriers for demand response at household customers. *Energy Procedia*, *75*, pp.1189-1196.

Vardakas, J.S., Zorba, N. & Verikoukis, C.V., 2015. A survey on demand response programs in smart grids: Pricing methods and optimization algorithms. *IEEE Communications Surveys & Tutorials*, *17*(1), pp.152-178.